In poetry class, I heard about "organic poetry" and "concrete poetry". My general understanding of the terms was that organic meant 'deriving form from the natural needs of the subject matter' whereas concrete meant something like 'painting the word "rock" on a rock and placing it somewhere where it could be read'.

I thought there should be "Organic Concrete" poetry, which might breathe more naturally than concrete poetry and might display form as a part of the substance and meaning of a poetic work. In the series "Adventures in AI Usage", of which this is the second installment -- "P(H)osers" being the first ,I try to use Firefly, Imogen, and Image GPT to come up with visual help defining and examining the concept of "Organic Poetry".

In this series, I let the reader see the prompts I used to come up with the images. I don't refine specific images, but rather let the AI generate new ones each time.. This second attempt to generate with AI did not come up with as many "can't do that" messages, like I did when I tried to put "cleavage" or "sexy" into the prompts for "P(H)osers". Each prompt comes in front of the image along with the name of the program that generated it. (Firefly 1,2,3, and 4 ultra, as well as, Imogen 3 and 4, and Image GPT)

piece of driftwood in the shape of a giant flower with the words "see the sea blooming" growing from the blossoms and florescent algae blooming in background

Firefly Image 4

piece of driftwood in the shape of a giant flower with the words "see the sea blooming" growing from the blossoms and florescent algae blooming in background

GPT

piece of driftwood in the shape of a giant flower with the words "see the sea blooming" growing from the blossoms and florescent algae blooming in background

imagen 3

piece of driftwood in the shape of a giant flower with the words "see the sea blooming" growing from the blossoms and florescent algae blooming in background

imogen 4

piece of driftwood in the shape of a giant flower with the words "see the sea blooming" growing from the blossoms and florescent algae blooming red in background

imogen 4

rock formation in the shape of a giant flower with the words "see the sea blooming" clearly visible and embossed in the blossoms and florescent algae blooming red in background

imogen 4

Google generated the following:

In literary terms, organic poetry draws its meaning and practice from the concept of "organic form" rather than a strict etymological root of the combined phrase "organic poetry".

Here's how to understand the etymology and development of the concept:

1. Organic

- The term "organic" itself originates from the Middle English period (1150-1500).
- It is partly borrowed from the French "organique" and the Latin "organum," which means "musical instrument" or "tool".
- In a broader sense, "organic" in literary criticism and aesthetics indicates a natural interrelationship between the parts of a work, where they seem to occur spontaneously rather than being artificially imposed, according to the Academy of American Poets.

2. Organic form

- The concept of organic form was developed by Samuel Taylor Coleridge in the early 19th century.
- He used this concept to argue that the works of William Shakespeare were not formless, but

rather had a structure that grew naturally from the author's subject and materials.

- Organic form stands in contrast to "mechanical form," which refers to works produced according to artificial rules.
- In organic form, the form and content are indivisibly fused, reflecting a natural growth, similar to that of a plant, says Wikipedia.

3. Organic poetry
    - The term "Organic" in relation to poetry is said to have come from correspondence between poets Robert Duncan and Denise Levertov in 1963.
    - Levertov, in her 1965 essay "Some Notes on Organic Form", defined it as a "method of apperception... of recognizing what we perceive, and is based on an intuition of an order, a form beyond forms... and of which man's creative works are analogies, resemblances, natural allegories".
    - Organic poetry focuses on a structure that emerges naturally from the poem's content, reflecting the interconnectedness found in the natural world.
    - This approach prioritizes the poet's intuition and feelings over adherence to traditional structures and rhyme schemes.

In essence, while "organic" has roots in terms for instruments and tools, its application to poetry, via the concept of "organic form," evolved to describe a creative process where the poem's structure develops naturally from its content and the poet's inner experience, rather than being imposed by external rules.

ORGANIC CONCRETE

I

Was

a

Splinter

In your ass
I tried to be
Don't like baby sitters

screaming tears that stay stuck

that's not worth anything;
the salivation, the mild discomfort
prod you when things go too smoothly
but what's the image? an organic pink?
an orange pain?
a single drop of your drool
tells more about you

but words

the mathematics of imagery
make themselves plain enough
if not inspired if not overdone
they fall from the tongue
like germ warfare and tickle
an image that will not laugh.
what's not worth anything? what do you want? --
answers? you want to talk about it?)

a sculpture of entwined rocks, papers and scissors with the words "Rock, Paper, Scissors" clearly readable

firefly ultra

a sculpture of squashed together rocks, papers and scissors with the words "Rock, Paper, Scissors" clearly readable

ultra

a sculpture of squashed together rocks, papers and scissors with the words "Rock, Paper, Scissors" clearly readable gpt image

a sculpture of squashed together rocks, papers and scissors with the words "Rock, Paper, Scissors" clearly readable. Imagen 3

a sculpture of squashed together rocks, papers and scissors with the words "Rock, Paper, Scissors" clearly readable. **Imagen 4**

a sculpture of squashed together rocks, papers and scissors all twisted around themselves with the words "Rock, Paper, Scissors" clearly readable

imagen 3

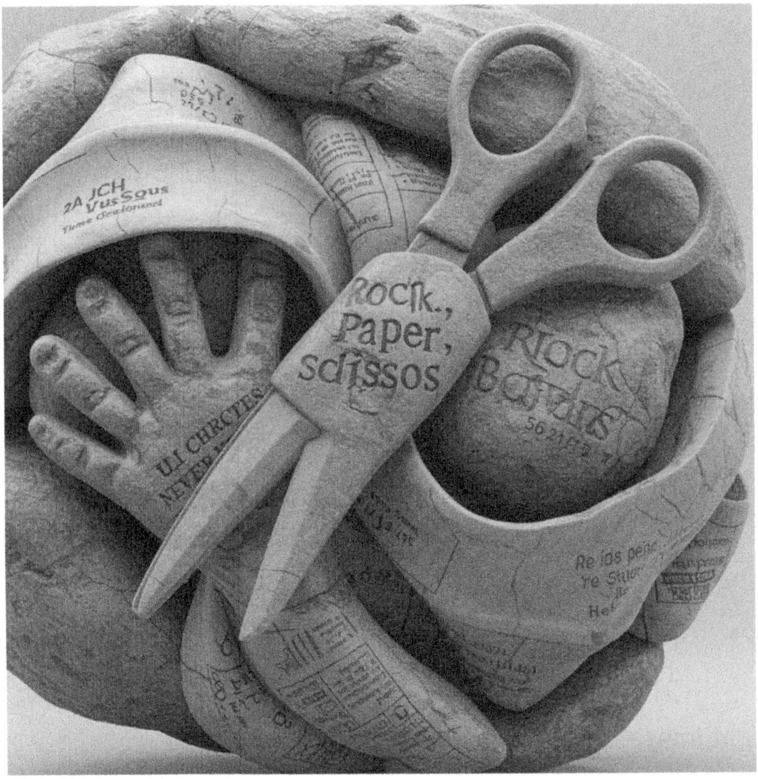

a sculpture of squashed together rocks, papers and scissors all twisted around themselves with the words "Rock, Paper, Scissors" clearly readable

gpt image

a colorful sculpture of squashed together rocks, papers and scissors all twisted around themselves with the words "Rock, Paper, Scissors" clearly readable with a background of adults partying. Gpt image

a colorful sculpture of squashed together rocks, papers and scissors all twisted around themselves with the words "Rock, Paper, Scissors" clearly readable with a background of adults partying. **Imogen 3**

a colorful sculpture of squashed together rocks, papers and scissors all twisted around themselves with the words "Rock, Paper, Scissors" clearly readable with a background of adults partying. Imogen 4

a colorful sculpture of squashed together rocks, papers and scissors all twisted around themselves with the words "Rock, Paper, Scissors" clearly readable with a background of adults partying.

Firefly image

a colorful sculpture of squashed together rocks, papers and scissors all twisted around themselves with the words "Rock, Paper, Scissors" clearly readable with a background of adults partying. Ferfly image 4 ultra

a colorful sculpture of squashed together rocks, papers and scissors all twisted around themselves with the words "Rock, Paper, Scissors" clearly readable with a background of adults partying. Firefly image 4

a colorful sculpture of squashed together rocks, papers and scissors all twisted around themselves with the words "Rock, Paper, Scissors" clearly readable with a background of a city being bombed. Gpt image

a sculpture of squashed together rocks, papers and scissors all twisted around themselves with the words "Rock, Paper, Scissors" clearly readable with a background of a city being bombed

gpt image

a sculpture of squashed together rocks, papers and scissors all twisted around themselves with the words "Rock, Paper, Scissors" clearly readable with a background of a city being bombed

imagen 3

a sculpture of squashed together rocks, papers and scissors all twisted around themselves with the words "Rock, Paper, Scissors" clearly readable with a background of a city being bombed

imogen 4

a sculpture of squashed together rocks, papers and scissors all twisted around themselves with the words "Rock, Paper, Scissors" clearly readable with a background of a city being bombed

firefly image 3

a sculpture of squashed together rocks, papers and scissors all twisted around themselves with the words "Rock, Paper, Scissors" clearly readable with a background of a city being bombed

firefly image 4

ROCK
PAPER
SCISSORS

a sculpture of squashed together rocks, papers and scissors all twisted around themselves with the words "Rock, Paper, Scissors" clearly readable with a background of a city being bombed

firefly image ultra

a colorful sculpture of squashed together rocks, papers and scissors all twisted around themselves with the words "Rock, Paper, Scissors" clearly readable with a background of a city being bombed

a colorful sculpture of squashed together rocks, papers and scissors all twisted around themselves with the words "Rock, Paper, Scissors" clearly readable with a background of a city being bombed

gpt image

a colorful sculpture of squashed together rocks, papers and scissors all twisted around themselves with the words "Rock, Paper, Scissors" clearly readable with a background of a city being bombed

imogen 3

a colorful sculpture of squashed together rocks, papers and scissors all twisted around themselves with the words "Rock, Paper, Scissors" clearly readable with a background of a city being bombed by jet bombers

gpt image

a colorful sculpture of squashed together rocks, papers and scissors all twisted around themselves with the words "Rock, Paper, Scissors" clearly readable with a background of a city being bombed by jet bombers

imogen 3

a colorful sculpture of squashed together rocks, papers and scissors all twisted around themselves with the words "Rock, Paper, Scissors" clearly readable with a background of a city being bombed by jet bombers

imogen 3

a colorful sculpture of squashed together rocks, papers and scissors all twisted around themselves with the words "Rock, Paper, Scissors" clearly readable with a background of a city being bombed by jet bombers

imogen 4

a colorful sculpture of squashed together rocks, papers and scissors all twisted around themselves with the words "Rock, Paper, Scissors" clearly readable with a background of a city being bombed by jet bombers

ultra

Looking at the very mixed results from "See the Sea Blooming" and "Rock, Paper, Scissors", I decided I'd better go for some Wikipedia explanations and clarifications. Here is what Wikipedia had to say about Concrete Poetry:

Concrete poetry is an arrangement of linguistic elements in which the typographical effect is more important in conveying meaning than verbal significance.[1] It is sometimes referred to as visual poetry, a term that has now developed a distinct meaning of its own. Concrete poetry relates more to the visual than to the verbal arts although there is a considerable overlap in the kind of product to which it refers. Historically, however, concrete poetry has developed from a long tradition of shaped or patterned poems in which the words are arranged in such a way as to depict their subject.

Development

Though the term 'concrete poetry' is modern, the idea of using letter arrangements to enhance the meaning of a poem is old. Such shaped poetry was popular in Greek Alexandria during the 3rd and 2nd centuries BCE, although only the handful which were collected together in the *Greek Anthology* now survive. Examples include poems by Simmias of Rhodes in the shape of an egg,[2] wings[3] and a hatchet,[4] as well as Theocritus' pan-pipes.[5]

The post-Classical revival of shaped poetry seems to begin with the *Gerechtigkeitsspirale* (spiral of justice), a relief carving of a poem at the pilgrimage church of St. Valentin, Kiedrich. The text is carved in the form of a spiral on the front of one of the church pews[6] and created in 1510 by master carpenter Erhart Falckener. But the heyday of the revival of shaped poetry came in the Baroque period when poets, in the words of Jeremy Adler, "did away with the more-or-less arbitrary appearance of the text, turned the incidental fact of writing into an essential facet of composition, and thereby...created a union of poetry with the visual arts".[7] There were already precedents for this in Micrography, a technique for creating visual images used by Hebrew artists, which involves organizing small arrangements of Biblical texts such that they form images which illustrate the subject of the text. Micrography allowed the creation of images of natural objects by Jews without directly breaking the prohibition of creating "graven images" that might be interpreted as idolatry. The technique is now used by both religious and secular artists and is similar to the use of Arabic texts in Islamic calligraphy.[8]

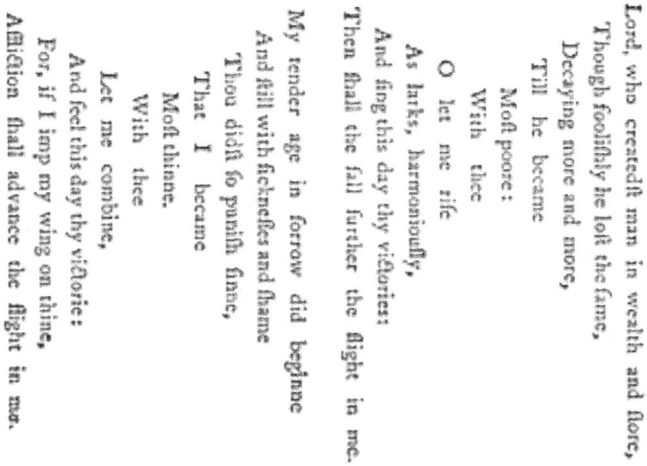
Lord, who createdst man in wealth and ſtore,
Though fooliſhly he loſt the ſame,
Decaying more and more,
Till he became
Moſt poore:
With thee
O let me riſe
As larks, harmoniouſly,
And ſing this day thy victories:
Then ſhall the fall further the flight in me.

My tender age in ſorrow did beginne
And ſtill with ſickneſſes and ſhame
Thou didſt ſo puniſh ſinne,
That I became
Moſt thinne.
With thee
Let me combine,
And feel this day thy victorie:
For, if I imp my wing on thine,
Affliction ſhall advance the flight in me.

George Herbert's "Easter Wings" (1633), printed sideways on facing pages so that the lines would call to mind angels flying with outstretched wings

Early religious examples of shaped poems in English include "Easter Wings" and "The Altar" in George Herbert's *The Temple* (1633)[9] and Robert Herrick's "This crosstree here", which is set in the shape of a cross, from his *Noble Numbers* (1647).[10] Secular examples include poems on the subject of drinking in the shape of wine flagons by Rabelais[11] and Charles-François Panard (1750),[12] supplemented by the elaborate goblet of Quirinus Moscherosch (1660),[13] the playful "A Toast" (Zdravljica, 1844) by France Prešeren, with stanzas in the shape of wine-glasses,[14] and "The Mouse's Tale", a shaped poem published in 1865 by Lewis Carroll.

The approach reappeared at the start of the 20th century, initially in the *Calligrammes* (1918) of Guillaume Apollinaire, with poems in the shape of a necktie, a fountain, and raindrops running down a window, among other examples. In that era also there were typographical experiments by members of avant-garde movements such as Futurism, Dada, and Surrealism in which layout moved from an auxiliary expression of meaning to artistic primacy. Thus the significance of the sound poetry in Marinetti's *Zang Tumb Tumb* (1912) is expressed through pictorial means. Similarly in Germany, Raoul Hausmann claimed that the typographic style of his "Phonemes" allowed the reader to recognize what sound was intended.[15] In Russia the Futurist poet Vasily Kamensky went so far as to term the typography of his *Tango with Cows*, published in 1914, "ferro-concrete poems" (*zhelezobetonnye poemy*), long before the name became current elsewhere.

A further move away from overt meaning occurred where "poems" were simplified to a simple arrangement of the letters of the alphabet. Louis Aragon, for example, exhibited the sequence from A to Z and titled it "Suicide" (1926), while Kurt Schwitters' "ZA (elementary)" has the alphabet in reverse,[16] and the Catalan writer Josep Maria Junoy (1885–1955) placed just the letters Z and A at the top and bottom of the page under the title "Ars Poetica".[17]

H.	Still wanting to have a THING that, with words on it, expresses something, some irony or cute juxtaposition, I continued into natural things made of concrete.

a vegatable garden with plants made of concrete

ultra

a vegetable planter with plants made of concrete

ultra

a vegetable planter with vegetables made of concrete

gpt

a vegetable planter with vegetables made of concrete

imogen 3

a vegetable planter with vegetables made of concrete

imogewn 4\\\

a vegetable planter with vegetables made of concrete

ultra

plastic apple with the words "sink you teeth into this" inscribed on the apple

ultra\\\

plastic apple with the words "sink your teeth into this" inscribed on the apple

gpt

imagen 3

imogen 4

Two ugly plastic pears with the words "Can't bear a pair of these" inscribed on the pears

Gpt

Imogen 3

Two ugly plastic pears with the words "Can't bear a pair of these" inscribed on the pears

mogen 4

Two ugly plastic pears with the words "Can't belmogen 4ar a pair of these" inscribed on the pearsand a bear in the background looking hungry

Two ugly plastic pears with the words "Can't bear a pair of these" inscribed on the pears and a bear in the background looking hungry

gpt

Two ugly plastic pears with the words "Can't bear a pair of these" inscribed on the pears on a stand with title " |A Pair of Pears"and a bear in the background looking hungry

Gpt

Dive deeper in AI Mode

AI responses may include mistakes. Learn more

Concrete poetry is a form of poetry where the visual arrangement of words on the page is as important as the words themselves, often creating a shape or image related to the poem's subject. The term "concrete poetry" itself has roots in the early 1950s, emerging simultaneously in Switzerland and Brazil, with Eugen Gomringer and the Noigandres group (Augusto de Campos, Haroldo de Campos, and Décio Pignatari) being key figures.

Elaboration:

- **Origins:**

  The term "concrete poetry" was coined independently in the early 1950s by Eugen Gomringer in Switzerland and the Noigandres group in Brazil. Gomringer, who had close ties with concrete art, is considered a founding figure of concrete poetry in the German-speaking world.

- **Meaning of "Concrete":**

  The term "concrete" in this context refers to the direct, tangible nature of the poetry. It emphasizes the physical form of the words on the page as an integral part of the poem's meaning, rather than just a vehicle for language.

- **Visual Element:**

  Concrete poetry prioritizes the visual arrangement of words, often creating shapes or images that relate to the poem's subject. This is in contrast to traditional poetry, where the visual presentation is typically secondary to the verbal content.

- **Historical Roots:**

  While the term is modern, concrete poetry has historical precedents in earlier forms like pattern poetry (e.g., George Herbert's "Easter Wings") and calligrams (e.g., Apollinaire's work).

- **International Movement:**

  Concrete poetry evolved into an international movement in the 1950s and 60s, with poets from various countries contributing to its development and exploration of its possibilities.

- **Key Characteristics:**
  Concrete poems often feature:
    - A strong relationship between the visual form and the poem's meaning.

- Intentional use of the visual aspects of language.
- A departure from traditional linear structure, emphasizing spatial arrangement.
- Potential for incorporating sound and performance, as exemplified by artists like Henri Chopin.

-
-
-

Who are some prominent concrete poets and examples of their work?

While Concrete Poetry encompasses a broad range of poets and styles, some notable figures and examples stand out for their significant contributions to the movement:

Key Figures:

- Eugen Gomringer: Often considered a pioneer of Concrete Poetry, Gomringer's work emphasizes visual minimalism and the arrangement of words to convey meaning. His poem *Silencio* uses the repeated word "silencio" to frame a blank space, representing the very idea of silence.

- Augusto de Campos and Haroldo de Campos: These Brazilian brothers were central figures in the Noigandres group, a key collective of Concrete Poets known for their innovative and experimental approach. Augusto de Campos's poem *Rever* (To See Again) explores the fusion of words and graphic design.
- Ian Hamilton Finlay: A Scottish poet, Finlay is known for his concrete poems often integrated into sculptures and garden landscapes, such as *The Present Order is the Disorder of the Future* installed in his garden, Little Sparta.
- Mary Ellen Solt: An American poet and critic, Solt's work often incorporates natural forms, like flowers, as seen in her poem *Forsythia*, where the letters of the word "forsythia" spring up to form the leaves of the plant.
- E. E. Cummings: While not strictly adhering to the core Concrete Poetry movement, Cummings' unconventional use of form, spacing, and typography in his poems,

like *Grasshopper*, served as a precursor and inspiration for Concrete Poets.

- Décio Pignatari: A member of the Noigandres group, Pignatari's poem *Beba Coca Cola* ("Drink Coca Cola") subverts advertising slogans through clever wordplay and visual arrangement, demonstrating the potential for social and cultural commentary within the form.
- Guillaume Apollinaire: Considered an early influence, Apollinaire's *Calligrammes* (1918) presented poems in the shapes of objects like a necktie or raindrops, foreshadowing the visual emphasis of Concrete Poetry.
- George Herbert: An earlier example of using form to enhance meaning can be seen in Herbert's 17th-century poem *Easter Wings*, where the lines are shaped like wings.

Other Notable Examples:

- Reinhard Döhl's "Apfel": This poem is famously shaped like an apple and consists primarily of the word "apple" and one instance of "worm".
- Lewis Carroll's "The Mouse's Tale": Found in *Alice's Adventures in Wonderland*, this poem

is arranged in the shape of a mouse's tail, playfully blending form and narrative.

These figures and their works demonstrate the diversity of approaches and themes within Concrete Poetry, highlighting the interplay between visual form and linguistic meaning that defines the movement.

AI responses may include mistakes. Learn more

What are some current trends in concrete poetry?
Are there other art movements related to concrete poetry?
Show me an example of Ian Hamilton Finlay's poem-sculptures

-
-
-

Show me an example of Ian Hamilton Finlay's poem-sculptures

One of Ian Hamilton Finlay's most well-known poem-sculptures, illustrating his unique approach to art and language, is
"The Present Order is the Disorder of the Future - Saint-Just" located in his garden, Little Sparta.

THE
REVOLUTION
ALL PRINCIPLES

RING
DOVE

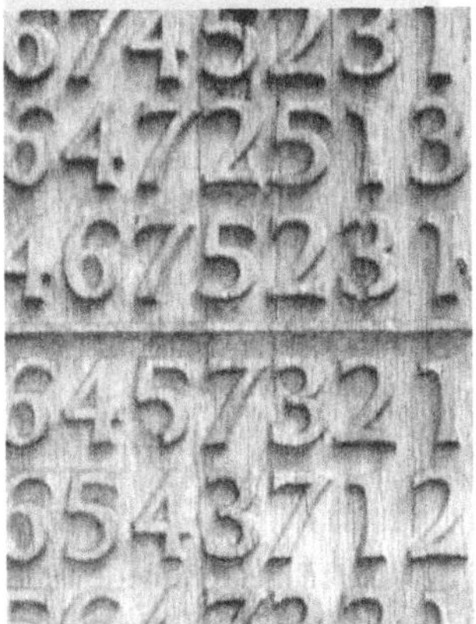
6745231
6472513
4675231
6457321
6543712

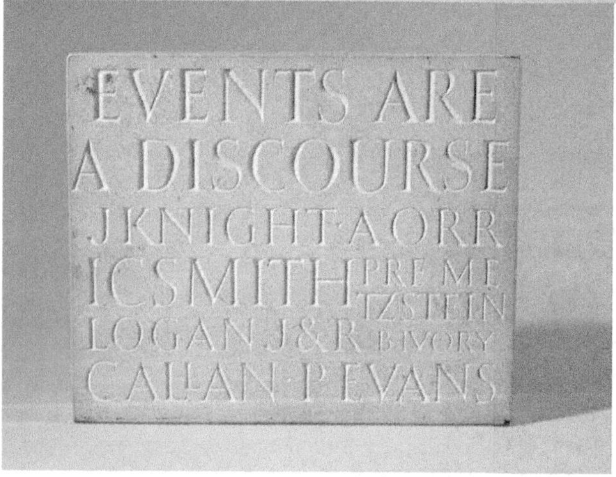

Here's a description of this work:

- Subject: This powerful piece consists of eleven large, rough-hewn stone blocks, arranged on the ground as if they were remnants of a fallen building.

- Text: Each stone is carved with a single word, forming the statement: "THE / PRESENT / ORDER / IS / THE / DISORDER / OF / THE / FUTURE / SAINT- / JUST".
- Meaning and Interpretation:
    - The phrase itself is attributed to Louis Antoine de Saint-Just, a prominent figure in the French Revolution.
    - The use of fragmented stone blocks suggests a ruin or the aftermath of destruction, perhaps alluding to the turbulent nature of revolution and societal change.
    - The careful, almost archaeological arrangement of the stones raises questions about the order of things and whether the current state of affairs is the intended one.
    - Finlay also created a print version of the work, suggesting that the viewer could rearrange the words, highlighting the potential for different interpretations and emphasizing the power of perspective.
    - The work's setting within the garden, described as a wild, "heathery plateau,"

further enhances its message of contrasting order and disorder, control and wildness.

- Impact: This sculpture challenges viewers to contemplate the nature of order and disorder, the fragility of established systems, and the potential for revolution and change. It encourages an active engagement with the piece and a questioning of societal norms.

In essence, "The Present Order is the Disorder of the Future - Saint-Just" is a powerful example of how Finlay blends sculpture, language, and landscape to create a work that is both visually striking and intellectually stimulating. It's a poem that extends beyond the confines of the page, becoming a tangible element within the environment and inviting contemplation on complex themes.

Elaborate on the French Revolution references in Finlay's work and their significance
What other artists explored similar themes of language and form?
Show me other examples of Finlay's work using Saint-Just quotes

-
-
-

What other artists explored similar themes of language and form?

Artists Exploring Language and Form in Art

Many artists have explored the intersection of language and visual form, often drawing inspiration from or extending the ideas of Concrete Poetry and Conceptual Art. Here are some prominent examples:

1. Conceptual Art:

- Joseph Kosuth: A leading figure in Conceptual Art, where the concept or idea behind the artwork is more important than the traditional art object itself. Works like *One and Three Chairs* use language to explore meaning and representation. The piece presents a chair, a photograph of the chair, and a dictionary definition of a chair. These elements prompt viewers to consider how objects are defined and understood through visual and linguistic means.
- Sol LeWitt: Another key figure in Conceptual Art, LeWitt focused on ideas and instructions for creating art, allowing others to realize his vision. He coined the term "conceptual art".

LeWitt's work often involves geometric forms and emphasizes the underlying concept. Instructions for creating wall drawings, for instance, highlight the importance of the idea over the artist's personal touch.

- Lawrence Weiner: Weiner is known for text-based artworks in the form of statements, often displayed on walls in public spaces. His work focuses on language and its potential to convey ideas directly. For example, he presents phrases such as "A 36" x 36" REMOVAL TO THE LATHING OR PLASTERING FROM A WALL".

2. Text and Image Combinations:

- Barbara Kruger: Kruger creates works that combine bold text with photographic images to critique societal norms, consumer culture, and power dynamics. Her signature style features black and white images overlaid with white-on-red text in a sans-serif font. She uses text as a direct and impactful tool to challenge viewers' assumptions.
- Jenny Holzer: Holzer utilizes text as the primary medium, presenting phrases on various surfaces, including billboards, buildings, and electronic displays.

Her *Truisms* series consists of one-line statements designed to provoke reflection and debate on social and political issues.

- Ed Ruscha: Ruscha's work features words and phrases within a visual context, exploring the interplay between language and image in popular culture and the urban landscape. He frequently uses typography as a visual element, transforming words into subjects of contemplation.

3. Other Related Approaches:

- The Art & Language Group: This influential collective explored the relationship between art and language through detailed discussions and textual works. They published a journal titled *Art-Language: The Journal of Conceptual Art* and created works that often consisted of text and conceptual ideas.
- Mel Bochner: Bochner combines language and visual elements in paintings with colorful words, exploring the visual aspects of language.

In Summary:

These are some examples of artists who have explored language and form in art. From the philosophical investigations of Conceptual Art to the social commentary of artists like Barbara Kruger and Jenny Holzer, the use of text in visual art continues to evolve and provoke dialogue.

Elaborate on the Art & Language group's approach to integrating language and art

What are some contemporary artists using text and image in their work?

Elaborate on how Jenny Holzer uses different mediums to display her text-based art

-
-
-

Elaborate on how Jenny Holzer uses different mediums to display her text-based art

Jenny Holzer's art is defined by its use of text and its strategic placement in diverse mediums, particularly public spaces

. She began with abstract painting but shifted towards language-based art, aiming to reach a wider audience beyond traditional art spaces. Holzer sees language as a powerful tool for conveying ideas and meanings directly.

Here are some of the key mediums she employs:

1. Printed Materials:

- Posters: Early in her career, Holzer anonymously displayed posters featuring her "Truisms" throughout Manhattan. These statements, such as "ABUSE OF POWER COMES AS NO SURPRISE" and "PROTECT ME FROM WHAT I WANT," challenged societal norms and encouraged reflection.

- Merchandise: Various merchandise items, including stickers, baseball caps, and t-shirts, have been used by Holzer to disseminate messages, allowing her art to reach a broader audience.
- Books and Publications: Holzer's texts have appeared in books and other printed formats, extending the reach of her work beyond installations and public spaces.

2. Electronic Displays:

- LED Signs: Holzer is known for using LED signs, which she began incorporating in the early 1980s. These electronic displays, often scrolling too fast to read easily, use the technology of public advertising and warning signs to deliver messages.
    - Public Locations: These signs have been installed in public spaces like Times Square and sports arenas, and on buildings and monuments.
    - Museum Installations: Holzer has created LED installations in museums, including a display at the Guggenheim Museum.

- Light Projections: Since the mid-1990s, Holzer has used light projectors to display texts onto buildings and landscapes worldwide. These projections often feature declassified government documents, bringing attention to issues of secrecy and transparency.

3. Stone and Architectural Elements:

   - Stone Benches and Plaques: Holzer engraves texts onto stone benches, plaques, and other architectural elements, creating works that blend into the environment but offer moments of reflection. These pieces can be found in public parks and communal spaces.
   - Marble Floors: For her Venice Biennale installation *The Child Room*, Holzer engraved a marble floor with her text, integrating her words into the architecture.

4. Other Platforms:

   - The Internet: Holzer has embraced digital platforms, participating in internet art projects like "Please Change Beliefs".
   - Oil Paintings: Holzer has returned to incorporating government documents, silk-screening them onto oil paintings to condemn brutality and military actions.

- Film and Video: Holzer has explored text-based art through film and video projects.

Holzer's diverse use of mediums allows the artist to:

- Reach a wider audience: By placing work in public spaces and using readily available materials, she engages individuals who might not typically visit art galleries.
- Create site-specific installations: Holzer tailors her work to the specific location and context, enhancing its meaning and impact.
- Provoke thought and discussion: Her texts, often presented in unexpected ways, prompt viewers to question their surroundings and consider different perspectives.
- Blur the lines between art and everyday life: By presenting her work in the street and using everyday objects, she challenges traditional notions of art and its place in society.

In summary, Jenny Holzer's use of various mediums to display her text-based art has made her a highly influential figure in contemporary art. She effectively combines language, form, and context to deliver messages and spark public dialogue.

Give examples of Holzer's light projections on buildings
What are some key themes in Holzer's texts beyond her Truisms?
How did Holzer's early painting background influence her text-based art?

-
-
-

What are some key themes in Holzer's texts beyond her Truisms?

Beyond her well-known "Truisms" series, Jenny Holzer explores a range of complex and often challenging themes in her text-based art, focusing on social justice, human rights, and the dynamics of power

.

Here are some of the key themes in her work:

1. Power and Politics:

- Critique of Authority: Holzer's art confronts issues of authority, abuse of power, and political corruption.
- Government Documents: She has extensively used declassified government documents, particularly related to military operations and

the treatment of detainees, to highlight issues of secrecy, transparency, and accountability.

- War and Violence: Holzer addresses the devastating consequences of war, both on individuals and on society, often incorporating personal accounts and testimony.
- Social and Political Injustice: Her work aims to expose social injustices, shed light on hidden truths, and provoke public discourse on pressing societal issues.

2. Feminism and Gender:

- Gender Identity and Equality: Holzer explores the complexities of gender identity, challenging societal norms and empowering individuals to question stereotypes and inequalities.
- Sexual Violence and Oppression: Her work addresses difficult subjects such as sexual assault against women and the power dynamics inherent in gender relationships.
- Women's Experiences and Challenges: Holzer uses her art to shed light on the experiences and challenges faced by women, advocating for gender equality and women's rights.

3. Human Rights and Social Justice:

- Freedom of Expression: Holzer champions freedom of expression and uses her art as a platform for dissent and critique of oppressive systems.
- Human Rights Initiatives: She collaborates with organizations like Amnesty International to raise awareness about human rights issues and advocate for social change.
- Social Justice Advocacy: Holzer actively supports non-profit organizations and grassroots movements working towards a more equitable society.

4. Language and Communication:

- The Power of Language: Language is a central element in Holzer's art, serving as a powerful tool to convey ideas, provoke thought, and challenge assumptions.
- Multiple Voices: She often uses a variety of voices, including her own, those of poets, politicians, and anonymous individuals, to represent diverse perspectives and highlight the subjectivity of truth.
- Bluring Boundaries: Holzer blurs the lines between public and private language, between

art and everyday life, encouraging viewers to engage with her work in unexpected ways.

In addition to these major themes, Holzer's work also touches on:

- Consumerism: Her work sometimes critiques consumer culture and the commodification of art.
- Survival: The "Survival Series" explores the anxieties and challenges of everyday life and the struggle for survival.
- Death and Grief: The "Laments" series reflects on death and grief, giving voice to those who have passed away.

Jenny Holzer's work, through its direct and thought-provoking language and its innovative use of various mediums, continues to challenge and engage audiences on critical social and political issues, extending far beyond the initial impact of her "Truisms."

Give an example of Holzer using declassified documents in her art

Elaborate on Holzer's use of anonymity in her early work

What other contemporary artists use text and language in their art?

-
-
-

Give an example of Holzer using declassified documents in her art

Jenny Holzer frequently uses declassified U.S. government documents in her art, particularly focusing on those related to military operations and the treatment of detainees in places like Afghanistan and Iraq

.

A notable example is the "Redaction Paintings" series, which began around 2005. In these works, Holzer meticulously reproduces declassified documents, often in silk-screened oil on linen, and highlights the redacted portions.

Here's how she uses these documents and what makes them impactful:

- Focus on the Missing: Instead of just presenting the readable text, Holzer draws attention to the blacked-out or redacted sections of the documents. These redactions, which often obscure crucial information related to sensitive intelligence or military methods, become a visual representation of secrecy and the denial of information to the public.
- Juxtaposition of Text and Image: Holzer often enlarges the documents, using bright colors to transform the black redaction blocks into abstract shapes. This transformation highlights the act of censorship itself and prompts viewers to consider the significance of what is being concealed.
- Symbolic Language: By emphasizing the redactions, Holzer creates a symbolic language that points to the government's efforts to control information and maintain a culture of secrecy.
- Varied Content: The documents used in the "Redaction Paintings" series include various materials such as memos, sworn statements, and reports related to military activities and the

treatment of prisoners. These documents often contain chilling details of violence, both legal and officially sanctioned.

- Commentary on Power and Control: Through her use of redacted documents, Holzer critiques the power structures that enable secrecy and questions the impact of information control on public discourse and understanding.

For example, in her painting "Top Secret 3" (2012), Holzer portrays a completely redacted "TOP SECRET" page from a Justice Department memorandum to a top CIA lawyer. Only the classification stamp and page number are visible, leaving viewers to ponder the nature of the suppressed information.

This series demonstrates Holzer's continued engagement with political themes and her innovative use of found texts to spark reflection on crucial issues like power, control, and government accountability.

AI Mode

CONCRETE VS PATTERN OR SHAPE POETRY. CONCRETE HAS MEANING IN WORDS TOO?

From Wikipedia, the free encyclopedia

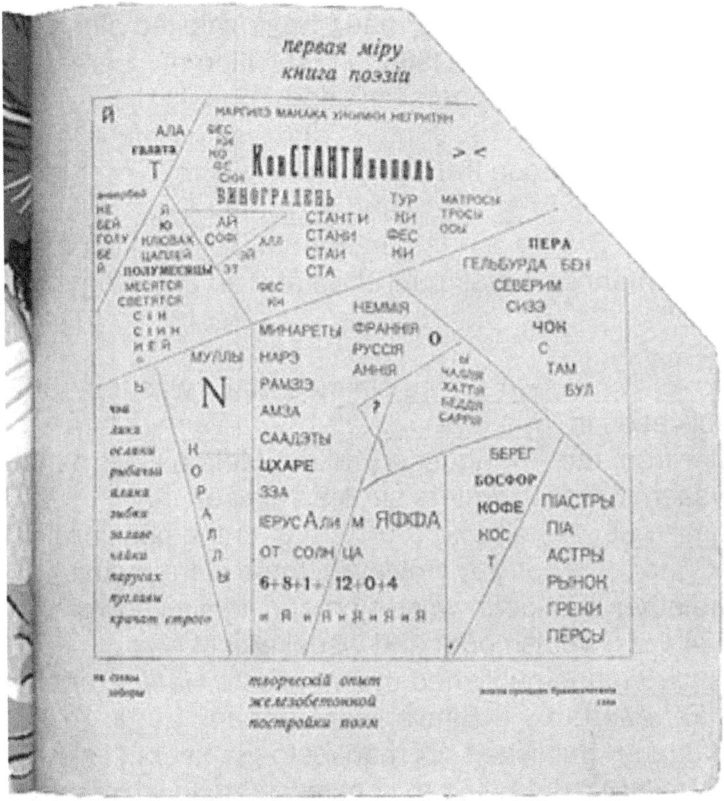

"Constantinople", a 'ferro-concrete poem' from *Tango with Cows* by the Russian Futurist Vasily Kamensky, 1914

Post-war concrete poetry

During the early 1950s two Brazilian artistic groups producing severely abstract and impersonal work were joined by poets linked to the São Paulo magazine *Noigandres* who began to treat language in an equally abstract way. Their work was termed "concrete poetry" after they exhibited along with the artists in the National Exhibition of Concrete Art (1956/57). The poets included Augusto de Campos, Haroldo de Campos and Décio Pignatari, who were joined in the exhibition by Ferreira Gullar, Ronaldo Azeredo and Wlademir Dias Pino from Rio de Janeiro. In 1958 a Brazilian concrete poetry manifesto was published and an anthology in 1962.[18]

Dom Sylvester Houédard claimed that it was the 1962 publication in *The Times Literary Supplement* of a letter from the Portuguese E.M. de Melo e Castro that awakened British writers such as himself, Ian Hamilton Finlay and Edwin Morgan to the possibilities of Concrete Poetry.[19] However, there were by this time other European writers producing similar work. In 1954 the Swedish poet and visual artist Öyvind Fahlström had published the manifesto *Hätila Ragulpr på Fåtskliaben.*[20][21] Similarly in Germany Eugen Gomringer published his manifesto *vom vers zur konstellation* (from line to constellation), in which he declared that a poem should be "a reality in itself" rather than a statement about reality,[22] and "as easily understood as signs in airports and traffic signs".[23] The difficulty in defining such a style is admitted by Houédard's statement that "a printed concrete poem is ambiguously both typographic-poetry and poetic-typography".[24]

Ian Hamilton Finlay sculpture in Stuttgart, 1975; the word *schiff* (ship) is carved in reverse and can only be decoded when reflected on water

Another difficulty of definition is caused by the way such works cross artistic boundaries into the areas of music and sculpture, or can alternatively be defined as sound poetry, visual poetry, found poetry and typewriter art.[25] Henri Chopin's work was related to his musical treatment of the word. Kenelm Cox (1927–68) was a kinetic artist "interested in the linear, serial aspects of visual experience but particularly in the process of change," whose revolving machines transcended the static page in being able to express this.[26] Ian Hamilton Finlay's concrete poetry began on the page but then moved increasingly towards three dimensional figuration and afterwards to site-specific art in the creation of his sculpture garden at Little Sparta. The Italian Maurizio Nannucci's *Dattilogrammmi* experiments (1964/1965) were also transitional, preluding his move into light art.[27]

Bob Cobbing, who was also a sound poet, had been experimenting with typewriter and duplicator since 1942. Of its possibilities in suggesting the physical dimension of the auditory process, he declared that "One can get the measure of a poem with the typewriter's accurate left/right & up & down movements; but superimposition by means of stencil and duplicator enable one to dance to this measure."[28] Houédard's entirely different work was also produced principally on the typewriter but approximates more to painterly and sculptural procedures.[29] So too does that of the American Minimalist artist Carl Andre, beginning from about 1958 and in parallel with his changing artistic procedures.[30] And in Italy Adriano Spatola (1941–88) developed the artistic fragmentation of language using various visual techniques in his *Zeroglifico* (1965/6).[31]

Edwin Morgan's experiments with concrete poetry covered several other aspects of it, including elements of found poetry 'discovered' by misreading and isolating elements from printed sources. "Most people have probably had the experience of scanning a newspaper page quickly and taking a message from it quite different from the intended one. I began looking deliberately for such hidden messages…preferably with the visual or typographical element part of the point."[32] Another aspect of the search for unintended concordances of meaning emerges in A Humument, the lifework of the visual artist Tom Phillips, who uses painterly and decorative procedures to isolate them on the page.

Despite such blurring of artistic boundaries, concrete poetry can be viewed as taking its place in a predominantly visual tradition stretching over more than two millennia that seeks to draw attention to the word in the space of the page, and to the spaces between words, as an aid to emphasising their significance.[33] In recent years, this approach has led Mario Petrucci to suggest that the "extreme example" of concrete poetry can be seen as nested within the larger concept of *Spatial Form*. Starting from the observation that poetry can usually be told from prose simply by looking at it, this reading of *Spatial Form* encompasses the many aspects of subtle visual significance that are held, for instance, by typeface or in the textures of repeated letters, as well as the more overt visual signals generated by the poem's layout.[34][35]

See also

- ASCII Art – Computer art form using text characters
- Calligram – Written work arranged in a way that creates a visual image
- Carmen figuratum – Poetic form
- List of concrete and visual poets
- musique concrete – Form of electroacoustic music
- Lyco art – Canadian painter and poet
- Something Else Press – Former American intermedia publisher
- Visual poetry – Poetry style that incorporates graphic and visual design elements

concrete poetry etymology

The term "concrete poetry" comes from the Latin word "concretus," meaning "condensed, hardened, stiff, curdled, congealed, clotted," and figuratively, "thick; dim."

It literally translates to "grown together," according to the Online Etymology Dictionary. The term itself is a translation of phrases coined in the mid-1950s in Brazil ("poesia concreta") and Germany ("die konkrete Dichtung").

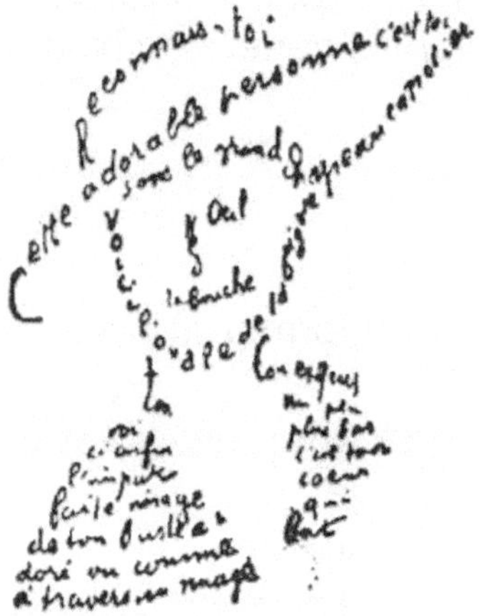

It's
cold outside.
I don't want to go
outdoors and play.
But mum says
I have to
anyway.
It's starting to snow
brrrr... and I'm going to freeze - brrrr...
I hate playing outside on days like these.
brrrr... But wait a sec, I've had the brrrr...
most amazing, brilliant idea!
I'll cover myself up
with snow and I'll
hide in here!

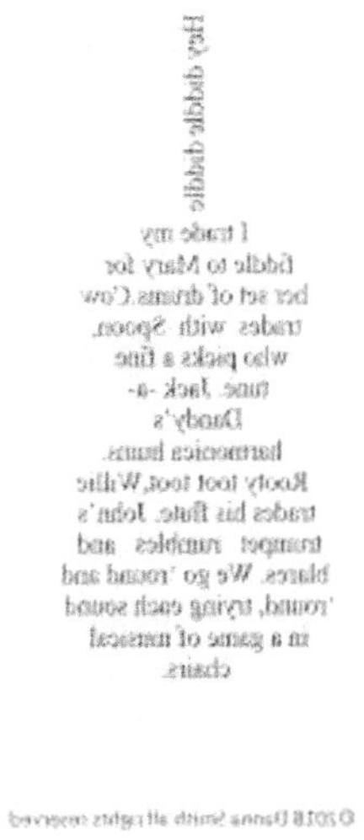

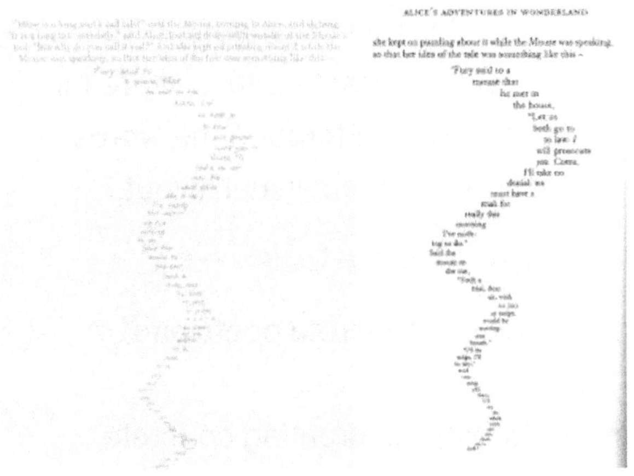

This etymological origin aligns with the core idea of concrete poetry: to "grow together" visual and verbal elements to create meaning. Concrete poetry emphasizes the visual and spatial aspects of language, using typography, layout, and other visual means to shape a poem's form and meaning. The visual arrangement of words is as important as the words themselves.

In simpler terms:

- Concrete: Referring to something tangible, specific, and real, rather than abstract.
- Poetry: A literary form that expresses ideas or emotions through imagery, sound, and rhythm.

Therefore, concrete poetry is a type of poetry that makes use of the physical arrangement of words on a page or in space to express its meaning, emphasizing the visual and tangible aspects of language. This approach contrasts with abstract poetry, where the meaning is primarily conveyed through the words themselves, independent of their visual layout.

Who are some prominent concrete poets and examples of their work?
What are some techniques in creating concrete poems?
What are some other artistic movements from the mid-20th century?

The search results suggest a possible confusion between two different poems:

- "Apfel" (German for "Apple") by Reinhard Döhl: This is a famous concrete poem written in 1965, shaped like an apple and made up primarily of the word "apple" and one instance of "worm". It is considered a seminal work in concrete poetry, according to Blogger.com.

The Deadend of Concrete Poetry

Reinhard Döhl, "Apfel" (1965)

One of the first concrete poems I can remember—one that I believe was up on Mrs Moomaugh's wall at the American

Cooperative School in Calacoto, Bolivia—was Reinhard Döhl's "Apfel" (which we could roughly translate as "Apple"). I believe this poem became the most influential concrete poem in our English class. When Mrs Moomaugh called on us to create our own concrete poems, each of our attempts was a riff off Döhl's oversimple trope: a poem both written out of and shaped like its subject.

We drew clouds full of "cloud," grass out of "grass," and trees from repetitions of "tree"—and those three were probably the most commonly encountered words in our poems. In a real way, Döhl's poem extracted our imaginations, because he showed us the easy way. In high school, everything can be frightening, but to suddenly be asked suddenly to be creative in the middle of the day can be one of the greatest shocks to children of that age. The metalinguistic depth of a simple (even stark) concrete poem like Eugen Gomringer's "silencio" escaped us. We had no idea how to recreate that poem. But Döhl, he showed us the way.

Döhl told us that all we needed to do with the words in a concrete poem was draw a picture—in his case, the uncompelling story of a worm inside an apple. He taught us that ideas were unimportant, that thought was ancillary, that the real object of poetry was a generous lack of depth.

And whenever I thought of Döhl's poem, even back when I lived in Gomringer's Bolivia, I found it lacking. I didn't like the shape of it. I didn't see the point of it. I thought what every visual-poetry-hating critic has thought: *There is no there there.* The poem taught us nothing and showed us little.

All of which makes me wonder why people frequently end up on this blog looking for a picture of this poem. What is it that draws people to this particular concrete poem? Maybe it is that the poem is easy enough for them. They can figure it out.

I see this all the time. People often tell me, with some glee, that they have figured out a visual poem of mine, as if it were nothing but a puzzle to be figured out, a code to be broken. When they tell me this, they have actually figured out a pun—verbal, visual, or hybrid—or discovered a word within the poem. But they are really holding a handful of dust out to me. They haven't figured out that the poem needs more than a password to open it. They haven't figured out that the poem is asking them to pause for a bit and dismantle it before they put it back together. They haven't figured out that the poem is a bit of verbo-visual language for their entertainment. They haven't figured out that the poem is always about joy and revelation.

ecr. l'inf.

Posted 23rd January 2006 by Geofhuth

What are examples of concrete poetry?

Secular examples include poems on the subject of drinking in the shape of wine flagons by Rabelais and Charles-François Panard (1750), supplemented by the elaborate goblet of Quirinus Moscherosch (1660), the playful "A Toast" (Zdravljica, 1844) by France Prešeren, with stanzas in the shape of wine-glasses, and "The ...

in the shape of an apple, generate words "apple, sib, apple, seb, apple tufaahatan apple maca apple"

firefly image 4

in the shape of an apple, generate words "apple, sib, apple, seb, apple tufaahatan apple maca apple"

gpt

in the shape of an apple, generate words "apple, sib, apple, seb, apple tufaahatan apple maca apple"

imogen 3

in the shape of an apple, generate words "apple, sib, apple, seb, apple tufaahatan apple maca apple"

imogen 4

in the shape of an pear, generate words "la pera, pear, Apion, pear, per, birnen, poire, كُمَّثرى , Apion , 梨子 . pear"

in the shape of an pear, generate words "la pera, pear, Apion, pear, per, birnen, poire, كُمَّثرَى , Apion , 梨子. pear"

gpt

in the shape of a concrete pear, generate words "la pera, pear, Apion, pear, per, birnen, poire, كُمَّثرَى , Apion , 梨子. pear"

gpt

in the shape of a concrete pear statue, generate words "la pera, pear, Apion, pear, per, birnen, poire, كُمّثرى , Apion , 梨子. pear"

gpt

in the shape of a concrete pear, generate words "la pera, pear, Apion, pear, per, birnen, poire, كُمّثرى , Apion , 梨子. pear". a statue with a town and people in the background

gpt

in the shape of a concrete pear, generate words "la pera, pear, Apion, pear, per, birnen, poire, گُمَّثری , Apion , 梨子. pear". a statue with a town and people in the background

ultra

a concrete tree statue in a park

ultra

a concrete tree statue in a park covered by the words "tree, tree, tree"

ultra

a concrete tree statue in a park covered by the words "tree, tree, tree"

gpt

a concrete tree statue in a park covered by the words "tree, tree, tree"

omgen 3

a concrete tree statue in a park covered by the words "tree, tree, tree"

imogen 4

rock of gibralter style statue in park with "rock" written all over it sitting next to a stone with "stone" written on it

imogen 4

marble slab underwater with "stone cold" inscribed on it

gpt

imogen3

imogen3

imogen4

TURTLE THAT LOOKS LIKE A GREEN BOX

Firefly image 4

TURTLE THAT LOOKS LIKE A GREEN BOX

Imogen4

puckered moon

imogen 4

IWow! Did I get sidetracked or what? I generated a monument that showed there was a difference between "Rock" and "Stone" but not necessarily in the best way. I kept at the concrete objects of nature but really don't know what I was thinking with the prompts that generated gravestones under the sea. "Box Turtle"!!! What kind of PUNishment was that?!!!! "Puckered moon?" Puckered Moon????!! I needed to refresh myself with the meaning and entymology of "Organic Poetry", so I went to google and this is what it said –

Organic Poetry

Paul E Nelson
https://paulenelson.com › organic-poetry

The term Organic is taken from correspondence between **Robert Duncan and Denise Levertov** from 1963 in which Robert Duncan described the difference between ...
Missing: ~~etymology~~ | Show results with: etymology

People also ask

What does organic mean in poetry?

What is the organic theory of poetry?

Where did the term "poetry" come from?

What are the three types of poetry?

"organic concrete" poetry etymology

"Organic concrete" poetry is not a recognized or common term in literary criticism or history

. It seems to be a combination of two distinct concepts:

- Organic Poetry: This refers to poetry where the structure and form emerge naturally from the content and the poet's intuition, rather than adhering to traditional rules like rhyme and meter. Organic form prioritizes the poem's emotional resonance and aims to reflect the natural rhythm of speech and the subject matter.
- Concrete Poetry: This type of poetry focuses on the visual arrangement of words and letters on the page, where the layout and design are as important as the words themselves in conveying meaning. Concrete poems often form a visual representation of their subject or theme.

Possible Interpretations:
Given the lack of a defined term "organic concrete" poetry, it's possible the user is trying to describe:

- Concrete poems that have an organic or natural form: Perhaps the poem's visual shape mirrors a natural object or pattern, or the arrangement of words evokes a sense of natural growth or movement, according to The Art of Education University.
- Concrete poems that feel spontaneous or less rigid: Concrete poetry can sometimes be very structured and abstract. A user might be contrasting this with concrete poems that feel more free-flowing or emotionally driven.

Etymology of "Concrete Poetry":
The term "concrete poetry" was coined in the early 1950s by European artists Max Bill and Öyvind Fahlström. The Brazilian group Noigandres also used the term and outlined its methods in their manifesto "Pilot Plan for Concrete Poetry". The movement gained popularity in the 1960s, with poets using words to form recognizable shapes and patterns. The word "concrete" itself comes from the Latin word "concretus," meaning "compact or condensed," according to Wikipedia. This root word reflects the solid and tangible nature of both building material and the visually organized language of concrete poetry.

AI responses may include mistakes. Learn more

Give me an example of a concrete poem with organic elements

What are other art movements exploring visual and textual elements besides organic and concrete poetry?

Who are some well-known organic poets?

-
-
-

What are other art movements exploring visual and textual elements besides organic and concrete poetry?

Absolutely! Besides "organic" and "concrete" approaches to poetry, many other art movements and styles explore the interplay between visual and textual elements. Here are some prominent examples:

- Lettrism

  : Founded by Isidore Isou in the 1940s, this movement focused on letters as basic units of poetic creation, prioritizing them as "sounds" and "images". They aimed to create purely formal poetry devoid of semantic content, utilizing techniques like hypergraphics, which merged writing with visual art.

- Conceptual Art

  : Conceptual art emphasizes the idea or concept behind the artwork over the final physical object. Text and imagery are frequently used to express these ideas, with language serving as an essential component of the artwork itself, according to Springer.

- Pop Art

  : Emerging in the 1950s, Pop Art challenged traditional art boundaries by incorporating imagery and elements from mass media and popular culture, like advertisements and comic books. It often features text-based elements, like words and phrases, alongside its visual imagery.

- Surrealism

  : This movement, known for its dreamlike and fantastical imagery, also experimented with the relationship between words and images. Surrealists used techniques like automatic writing and collage to explore the subconscious mind and create works that disrupted conventional narratives.

- Dadaism

  : Preceding Surrealism, Dadaists also explored the merging of visual and textual forms, often incorporating written and spoken words into their eclectic mixed media works, according to metafora-studio-arts.org. Dadaists utilized collage extensively, using a wide array of iconography and materials to challenge traditional art perceptions, according to My Modern Met.

- Visual Poetry

  : This broad term encompasses various approaches that combine poetry with visual art elements. Visual poetry uses the layout and appearance of words, letters, and symbols to create meaning and enhance the emotional impact of the text. Concrete poetry and calligrams (where text forms a picture) are examples of visual poetry.

These movements demonstrate the diverse ways artists have merged visual and textual elements to challenge artistic conventions and explore new forms of expression.

AI responses may include mistakes. Learn more

What are some contemporary artists working with text and image?

How did Dada and Surrealism differ in using text and image?

What are the key differences between Lettrism and Concrete Poetry?

-
-
-

What are the key differences between Lettrism and Concrete Poetry?

While both Lettrism and Concrete Poetry explore the visual and textual aspects of language, they have some key differences in their focus and approach:

Lettrism:

- Focus on the Letter as a Basic Unit: Lettrism, as its name suggests, emphasizes the letter as the fundamental building block of poetic and artistic creation. Isidore Isou, the founder, aimed to deconstruct poetry to its most basic element, the individual letter, seeing it as both a visual symbol and a potential sound.
- Emphasis on Hypergraphics: Lettrists developed techniques like hypergraphics,

which involved the merging of writing and visual art, utilizing a wide range of signs and symbols beyond just conventional letters, according to Tumblr.

- Experimentation with Sound: Lettrism also delved into the auditory potential of letters and language, incorporating elements of sound poetry and performance art.
- Rejection of Semantic Coherence: Some Lettrist works deliberately discounted semantic and syntactical coherence, aiming for a purely formal poetry that prioritized the visual and auditory experience of the letter forms.
- Influence on Diverse Fields: Lettrism's influence extended beyond poetry and visual arts to encompass film, architecture, and even political theory.

Concrete Poetry:

- Focus on the Arrangement of Words: Concrete poetry, while also utilizing visual elements, centers around the typographical arrangement of words and phrases, where the layout and visual appearance significantly contribute to the poem's meaning.
- Visual Representation of Meaning: Concrete poems often form a visual image that reflects

or comments on the poem's subject or theme, with the words or lines of the poem themselves creating a picture.

- Integration of Language and Visuals: Concrete poetry aims for a close relationship between the language and the visuals, with the layout and appearance of words and symbols enhancing the poem's message.
- Exploration of Ideograms: The movement was often accompanied by an interest in ideograms, considering words themselves as potential ideograms and exploring typographical innovations.
- Recognition as Poetry: While concrete poetry incorporates visual elements, it is still generally recognized as a form of poetry composed of typographical elements.

In summary:
Lettrism deconstructs poetry to the fundamental level of the letter, exploring its visual and auditory aspects, while Concrete Poetry focuses on the visual arrangement of words and phrases to create a visually representative form that enhances the poem's meaning.

IMAGE THAT EXPLORES THE CONCEPT OF "ORGANIC CONCRETE", ESPECIALLY WHEN IT COMES TO ORGANIC POETRY AND CONCRETE POETRY

Firefly image 4

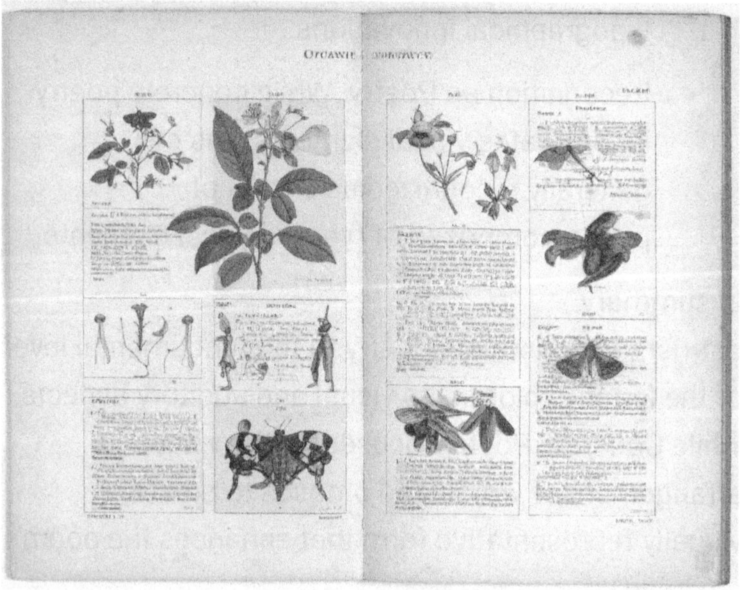

IMAGE THAT EXPLORES THE CONCEPT OF "ORGANIC CONCRETE", ESPECIALLY WHEN IT COMES TO ORGANIC POETRY AND CONCRETE POETRY

Gpt

IMAGE THAT EXPLORES THE CONCEPT OF "ORGANIC CONCRETE", ESPECIALLY WHEN IT COMES TO ORGANIC POETRY AND CONCRETE POETRY

Imogen3

IMAGE THAT EXPLORES THE CONCEPT OF "ORGANIC CONCRETE", ESPECIALLY WHEN IT COMES TO ORGANIC POETRY AND CONCRETE POETRY

Imagen 3

IMAGE THAT EXPLORES THE CONCEPT OF "ORGANIC CONCRETE", ESPECIALLY WHEN IT COMES TO ORGANIC POETRY AND CONCRETE POETRY

Imogen 3

IMAGE THAT EXPLORES THE CONCEPT OF "ORGANIC CONCRETE", ESPECIALLY WHEN IT COMES TO ORGANIC POETRY AND CONCRETE POETRY**Imogen4**

IMAGE THAT EXPLORES THE CONCEPT OF "ORGANIC CONCRETE", ESPECIALLY WHEN IT COMES TO ORGANIC POETRY AND CONCRETE POETRY

Imogen 4

I

Gpt

IMAGE THAT EXPLORES THE CONCEPT OF "ORGANIC CONCRETE", ESPECIALLY WHEN IT COMES TO ORGANIC POETRY AND CONCRETE POETRY

Hpt

IMAGE THAT EXPLORES THE CONCEPT OF "ORGANIC CONCRETE", ESPECIALLY WHEN IT COMES TO ORGANIC POETRY AND CONCRETE POETRY

Imogen3

IMAGE THAT EXPLORES THE CONCEPT OF "ORGANIC CONCRETE", ESPECIALLY WHEN IT COMES TO ORGANIC POETRY AND CONCRETE POETRY

Imogen3

IMAGE THAT EXPLORES THE CONCEPT OF "ORGANIC CONCRETE", ESPECIALLY WHEN IT COMES TO ORGANIC POETRY AND CONCRETE POETRY

Imogen 3

IMAGE THAT EXPLORES THE CONCEPT OF "ORGANIC CONCRETE", ESPECIALLY WHEN IT COMES TO ORGANIC POETRY AND CONCRETE POETRY

IMAGE THAT EXPLORES THE CONCEPT OF "ORGANIC CONCRETE", ESPECIALLY WHEN IT COMES TO ORGANIC POETRY AND CONCRETE POETRY

Imogen 4

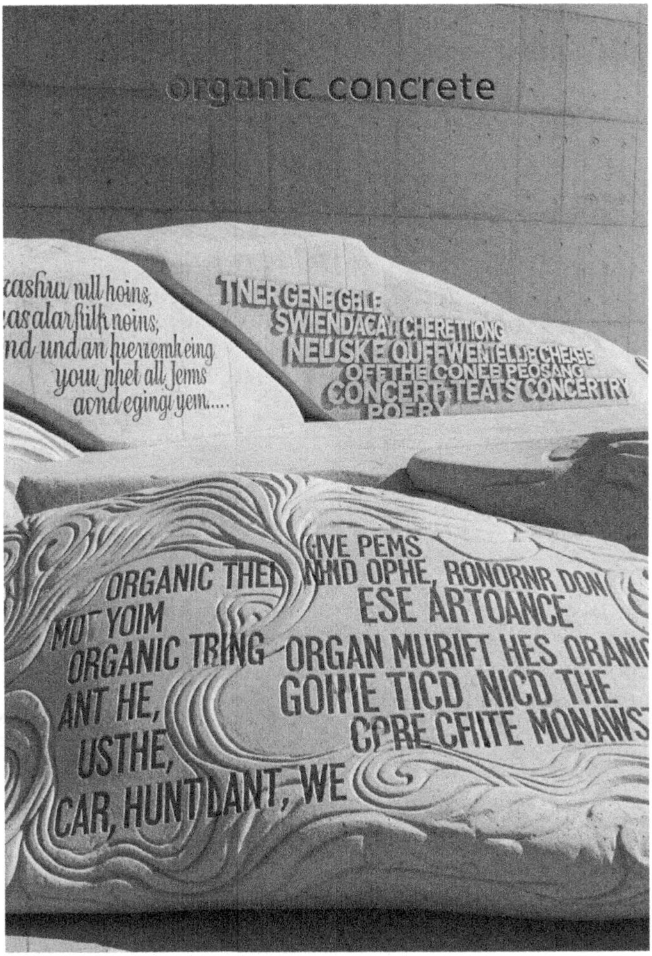

IMAGE THAT EXPLORES THE CONCEPT OF "ORGANIC CONCRETE", ESPECIALLY WHEN IT COMES TO ORGANIC POETRY AND CONCRETE POETRY

Ultra

IMAGE THAT EXPLORES THE CONCEPT OF "ORGANIC CONCRETE", ESPECIALLY WHEN IT COMES TO ORGANIC POETRY AND CONCRETE POETRY ABOUT CONCRETE AND ORGANIC

Imogen 4

IMAGE THAT EXPLORES THE CONCEPT OF "ORGANIC CONCRETE", ESPECIALLY WHEN IT COMES TO ORGANIC POETRY AND CONCRETE POETRY ABOUT CONCRETE AND ORGANIC

Gpt

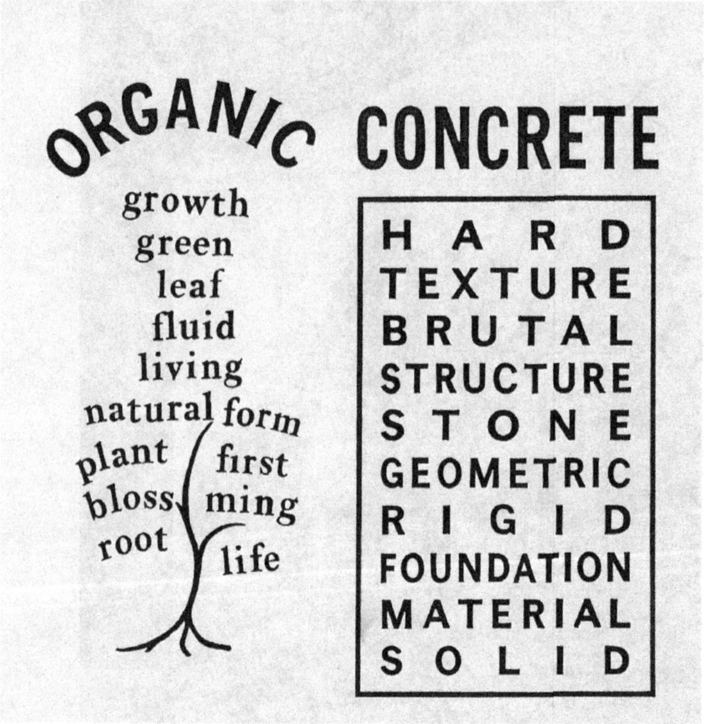

IMAGE THAT EXPLORES THE CONCEPT OF "ORGANIC CONCRETE", ESPECIALLY WHEN IT COMES TO ORGANIC POETRY AND CONCRETE POETRY ABOUT CONCRETE AND ORGANIC

Gpt

IMAGE THAT EXPLORES THE CONCEPT OF "ORGANIC CONCRETE", ESPECIALLY WHEN IT COMES TO ORGANIC POETRY AND CONCRETE POETRY ABOUT "CONCRETE" AND "ORGANIC"

Gpt

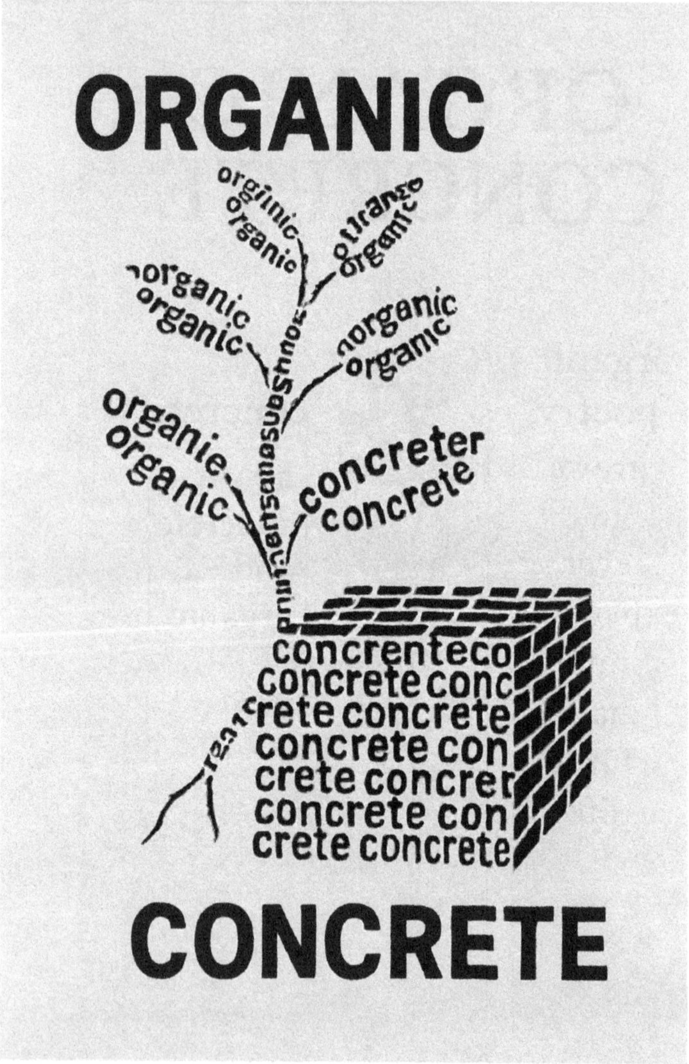

IMAGE THAT EXPLORES THE CONCEPT OF "ORGANIC CONCRETE", ESPECIALLY WHEN IT COMES TO ORGANIC POETRY AND CONCRETE POETRY ABOUT CONCRETE AND ORGANIC

Gpt

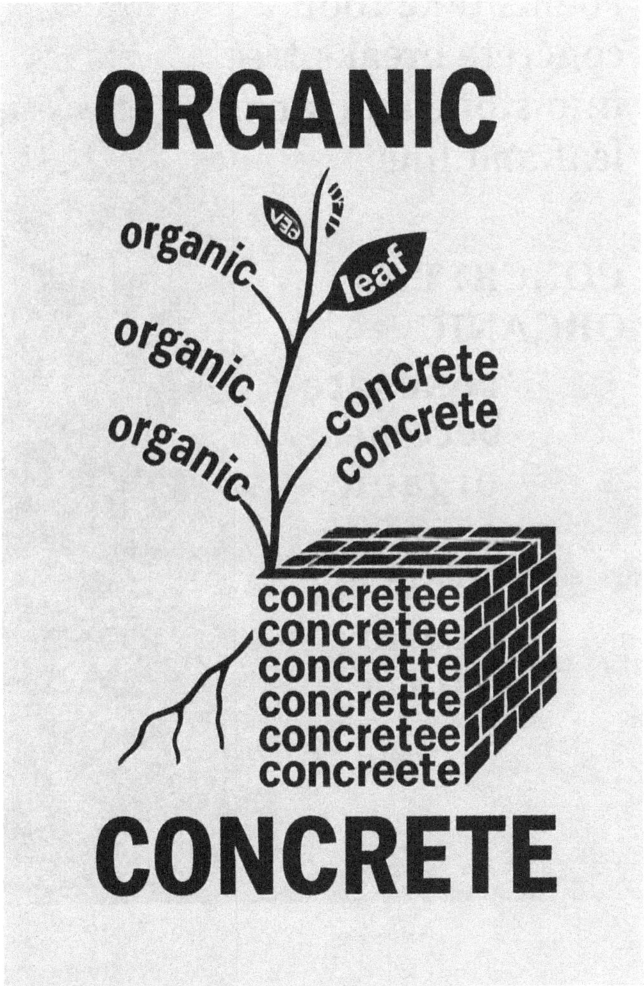

ORGANIC
CONCRETE

Poems take root
concrete breaks free
into stone and stem
leaf and line

CONCRITE
ORGANIC

concrete
becomes
organic
as words
grow in
solid
form

IMAGE THAT EXPLORES THE CONCEPT OF "ORGANIC CONCRETE", ESPECIALLY WHEN IT COMES TO ORGANIC POETRY AND CONCRETE POETRY ABOUT CONCRETE AND ORGANIC

Gpt

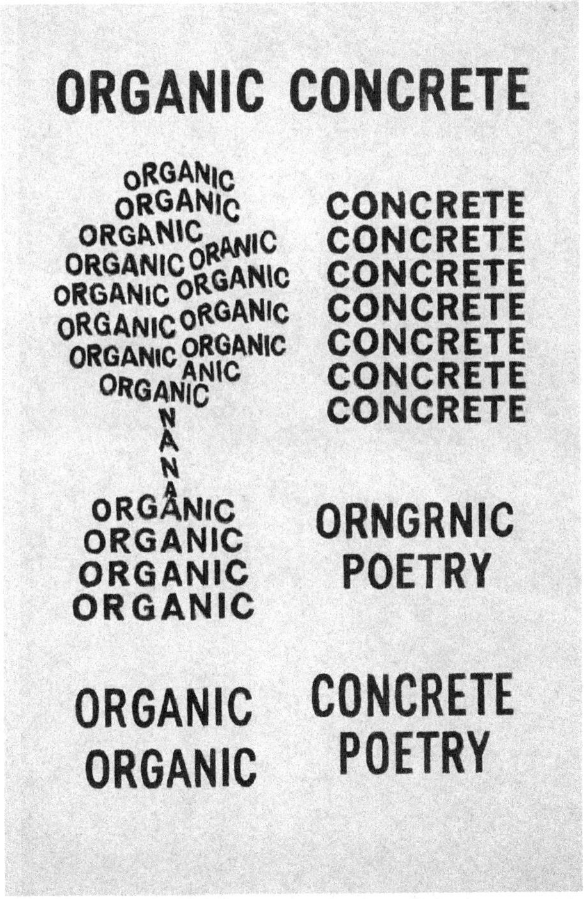

IMAGE THAT EXPLORES THE CONCEPT OF "ORGANIC CONCRETE", ESPECIALLY WHEN IT COMES TO ORGANIC POETRY AND CONCRETE POETRY ABOUT CONCRETE AND ORGANIC

Imogen 3

IMAGE THAT EXPLORES THE CONCEPT OF "ORGANIC CONCRETE", ESPECIALLY WHEN IT COMES TO ORGANIC POETRY AND CONCRETE POETRY ABOUT "CONCRETE" AND "ORGANIC"

Gpt

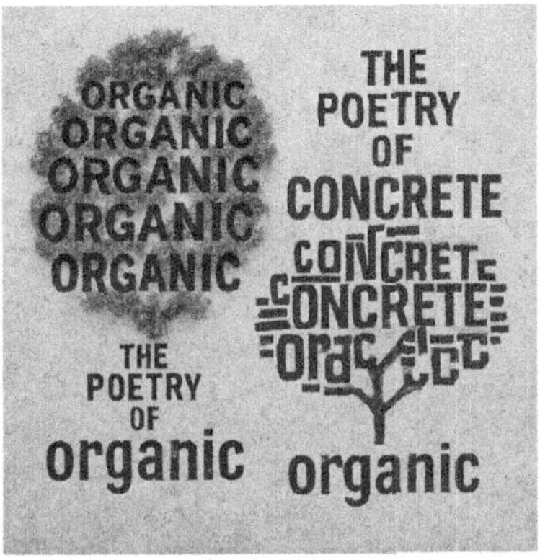

IMAGE THAT EXPLORES THE CONCEPT OF "ORGANIC CONCRETE", ESPECIALLY WHEN IT COMES TO ORGANIC POETRY AND CONCRETE POETRY ABOUT "CONCRETE" AND "ORGANIC"

Imogen 4

IMAGE THAT EXPLORES THE CONCEPT OF "ORGANIC CONCRETE", ESPECIALLY WHEN IT COMES TO ORGANIC POETRY AND CONCRETE POETRY ABOUT "CONCRETE" AND "ORGANIC"

Ultra

IMAGE THAT EXPLORES THE CONCEPT OF "ORGANIC CONCRETE", ESPECIALLY WHEN IT COMES TO ORGANIC POETRY AND CONCRETE POETRY ABOUT "CONCRETE" AND "ORGANIC"

Ulrra

IMAGE THAT EXPLORES THE CONCEPT OF "ORGANIC CONCRETE", ESPECIALLY WHEN IT COMES TO ORGANIC POETRY AND CONCRETE POETRY ABOUT CONCRETE AND ORGANIC and makes poetry

Gpt

\

MAGE THAT EXPLORES THE CONCEPT OF "ORGANIC CONCRETE", ESPECIALLY WHEN IT COMES TO ORGANIC POETRY AND CONCRETE POETRY ABOUT CONCRETE AND ORGANIC and sample of what is "organic concrete poetry"

Gpt

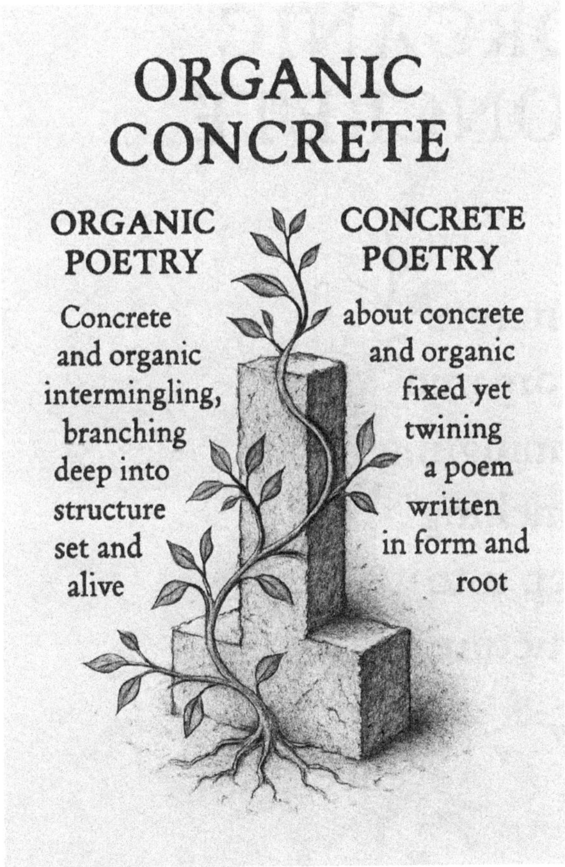

iMAGE THAT EXPLORES THE CONCEPT OF "ORGANIC CONCRETE", ESPECIALLY WHEN IT COMES TO ORGANIC POETRY AND CONCRETE POETRY ABOUT CONCRETE AND ORGANIC and sample of what is "organic concrete poetry"i

gpt

IMAGE THAT EXPLORES THE CONCEPT OF "ORGANIC CONCRETE", ESPECIALLY WHEN IT COMES TO ORGANIC POETRY AND CONCRETE POETRY ABOUT CONCRETE AND ORGANIC and sample of what is "organic concrete poetry"

Imogen 4

organic concrete

IMAGE THAT EXPLORES THE CONCEPT OF "ORGANIC CONCRETE", ESPECIALLY WHEN IT COMES TO ORGANIC POETRY AND CONCRETE POETRY ABOUT CONCRETE AND ORGANIC and sample of what is "organic concrete poetry"

Imogen 4

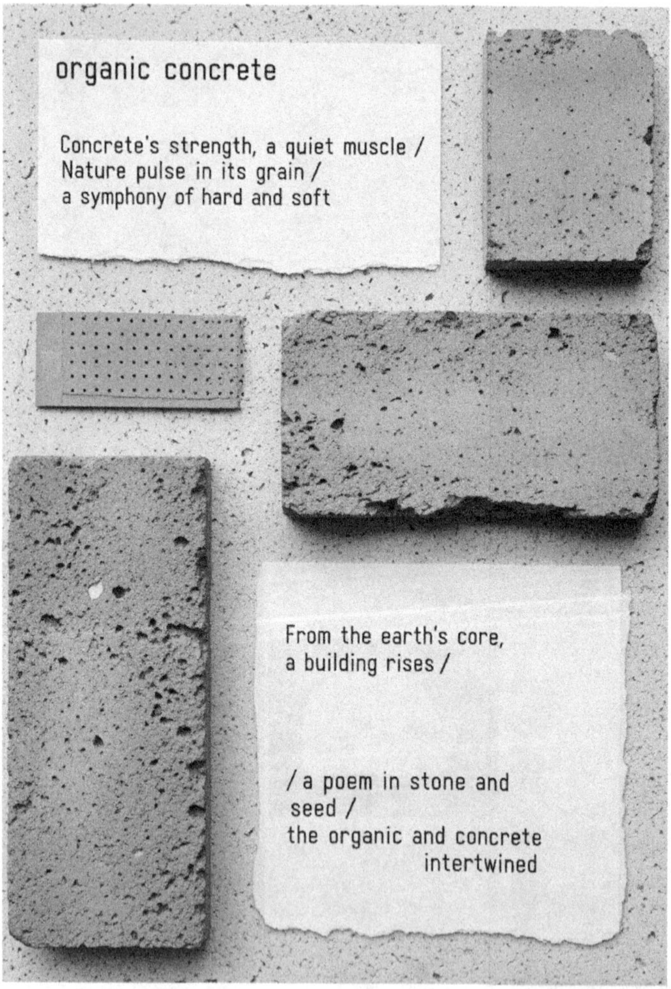

IMAGE THAT EXPLORES THE CONCEPT OF "ORGANIC CONCRETE", ESPECIALLY WHEN IT COMES TO ORGANIC POETRY AND CONCRETE POETRY ABOUT CONCRETE AND ORGANIC and sample of what is "organic concrete poetry"

Imogen 3

IMAGE THAT EXPLORES THE CONCEPT OF "ORGANIC CONCRETE", ESPECIALLY WHEN IT COMES TO ORGANIC POETRY AND CONCRETE POETRY ABOUT CONCRETE AND ORGANIC and sample of what is "organic concrete poetry"

Ultra

GENERATE POETRY THAT COULD BE DESCRIBED AS "ORGANIC CONCRETE POETRY"

Ultra

GENERATE WORDS AND IMAGE POETRY THAT COULD BE DESCRIBED AS "ORGANIC CONCRETE POETRY"

Ultra

POETRY THAT COULD BE CALLED "ORGANIC CONCRETE POETRY"

Ultra

Made in United States
North Haven, CT
05 November 2026

21789891R00085